CUTTY SARK

Of all her lovely sisters that roamed the seven seas
in the golden age of sail, only the Cutty Sark remains
to show the world in which we live what a thing of
grace and beauty was a clipper ship
in the fullness of her glory.

FRANK CLARK

CUTTY SARK
OFF No.
63557
1869
LONDON

The Great Ship Race

In the middle years of the 19th century, the annual race of tea clippers from China was as much an obsession for seafarers as these days the Grand National is to the racing man. Huge sums often changed hands in bets, and crews would risk their wages as well as their lives.

Each winter for over 20 years saw tea clippers sail for the Orient to race back with the first of the new season's tea. These clippers were the greyhounds of the ocean, built for speed – ships whose very names breathed the spirit of romance: names such as *Ariel*, *Taeping*, *Titania*, *Sir Lancelot* and *Belted Will*.

The winning owner would receive a small fortune in stakes, and the captain, besides acquiring a tidy sum of money, would have his reputation made for life.

The outward run from London would take general cargo anywhere in the East or perhaps Australia. The ships would then head for China, perhaps doing an intermediate coastal run, for example from Bangkok to Hong Kong with a cargo of rice. Towards high summer, the clippers would make their way to Shanghai or Foochow, China's main tea ports, in readiness to receive the first of the season's crop.

Even after the opening of the Suez Canal in 1869 enabled steamships to compete in the tea trade, owners of sailing ships remained confident. Many tea merchants thought that tea travelled better in wooden hulls than in iron. In any case a sharp increase in trade promised that there would be work for ships of all sorts.

The scene on the river at Shanghai was a wonderful one, with men-o'-war at anchor with their awnings out, the bustle of a thousand junks and sampans, and tiny gigs flitting here and there taking

BELOW:
Three crew members, including (right) Tony Robson, the celebrated Chinese cook who, as a baby, had been found by a British ship alone on a raft in mid-ocean.

BOTTOM:
The Cutty Sark *off Shanghai, painted by Gerhard Geidel.*

The Great Ship Race

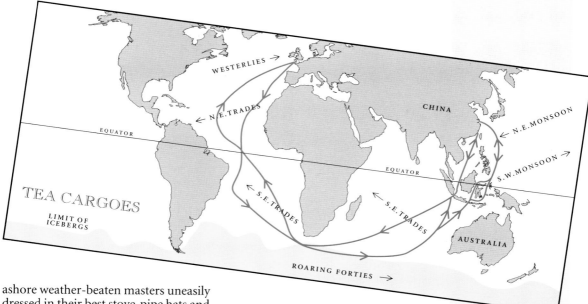

ABOVE:
The tea route to and from the Far East. In Pepys's day, the tea was carried by plodding East Indiamen, their monopoly protected by the Navigation Acts. The repeal of these Acts in 1849 allowed the faster American cotton clippers to enter the tea trade. These commanded double rates because of their speed. British rivals inevitably followed. The port of Aberdeen built the finest vessels. So brilliantly was the design developed and so daringly were the ships sailed that by 1860 only British ships were left in the race.

ashore weather-beaten masters uneasily dressed in their best stove-pipe hats and black cloth coats. Snatches of sea shanties from the merchantmen rang out across the water.

Presiding over the whole scene were the stately tea clippers with their sparkling brasswork and gleaming decks. Such was the pride in these vessels that a good officer would consider himself disgraced if in harbour even the slightest bit of rope was out of place, particularly if the berth was next to a Yankee ship.

For these clippers, there was often a frustrating period of waiting for a charter at the right price, around £3 per 50 cubic feet (1.41m³) of space, and a ship might even sail to a different port to find an acceptable deal. To add to the frustration, the new tea was slow to arrive from the interior, and there was rivalry amongst the top-line ships as to which could load first.

Once everything was agreed and the tea loaded, the ship would be towed out to sea, all flags flying, to the sound of saluting gunfire and the friendly cheers of the other crews. The race was on, and the winner would be the first ship to dock in London, some 100 days later.

Every shipowner, most of them ex-masters themselves, had just one ambition – to win this 'Great Ship Race', as the newspapers called it. One such man was Bernard Waymouth. To capture what was then the Blue Riband of the sea, he had *Thermopylae* specially built at Aberdeen in 1868. Even on her maiden voyage she covered the trip to Melbourne in 60 days, a record that has never been beaten. Waymouth gave her a gold cock to wear on her mainmast as the sign of her supremacy.

It was to beat this ship of all others that the *Cutty Sark* was built.

CUTTY SARK

'I never sailed a finer ship. At 10 or 12 knots she did not disturb the water at all. She was the fastest ship of her day, a grand ship, and a ship that will last for ever.

CAPTAIN GEORGE MOODIE
FIRST MASTER OF THE *CUTTY SARK*

Building the Cutty Sark

ABOVE:
The registration certificate of the Cutty Sark, *still held by the Registry of Shipping, Cardiff.*

FAR RIGHT:
Hercules Linton, the young designer of the Cutty Sark, *whose achievement lay in copying* The Tweed's *bow while redesigning the stern and the bottom. However, the building of the* Cutty Sark *was to sink his firm.*

RIGHT:
John Willis, known as 'White Hat' Willis, the fleet owner whose determination to win the Great Ship Race led to the Cutty Sark *being built.*

Many owners were repeatedly disappointed with their ships' performance in the race. One such was John 'Jock' Willis, an old sailing ship master who had 'swallowed the anchor' and settled in London to manage his fleet. His heart's desire was to win the race although his previous first-string vessels, *Lammermuir* in the 50s and *Whiteadder* in the 60s had been outclassed by the opposition.

But Willis was not a man to give in easily. He had the seeds of a design in his head, for his ship *The Tweed*, he reckoned, was the fastest ship afloat but was too big for the tea trade. If he could build something with the same lines, but smaller . . .

He had just the man to do it, a talented young designer called Hercules Linton, a partner in a new Glasgow firm, Scott & Linton.

In 1868, Willis took Linton to see *The Tweed* and a plan was formed. The new ship would have *The Tweed*'s bow, not too rounded at the forefoot (the front of the keel) but with a squarer stern than the bigger ship. This was a telling factor in the *Cutty Sark*'s ultimate success, for it made her more stable than *The Tweed* and more able to carry sail. Most tea clippers needed very careful handling in heavy seas, but the *Cutty Sark* could be driven hard like no other. The midship section owed its design to the Firth of Forth fishing boats, much admired by Linton, whose genius showed itself in the way he moulded the features together to create a beautiful new design.

Willis was a canny Scot. He wanted the best ship for the least possible outlay. Scott & Linton had never built anything as big as a clipper ship before, and as this would be a prestige job, Willis was able to

pressure them into quoting a figure of only £17 per ton. The contract price for the *Cutty Sark* amounted to no more than £16,150, a figure so low that it would ultimately break the firm that built her before the ship was complete. The final details of the fitting-out were completed by William Denny & Brothers.

The *Cutty Sark*'s decks were of solid teak, compared with *Thermopylae*'s pine, and she had 'tween decks laid, where

BUILDING THE CUTTY SARK

Thermopylae had none. This gave her a rigidity which other ships had not, and could account for a fact mentioned in Lubbock's book *The Log of the Cutty Sark* – that the *Cutty Sark* survived 'with a tight bottom, whilst the *Thermopylae* lies fathoms deep off the coast of Portugal'.

The *Cutty Sark* was launched on 22 November 1869 at Dumbarton. She was 'all ship', as seafarers used to say – with lovely sharp lines and perfect balance. Just the look of her at launch gave a foretaste of the speed and power she was soon to produce.

'CUTTY SARK'

The curious name 'Cutty Sark' is taken from the Robert Burns poem *Tam O'Shanter*. This recalls the ancient legend of Tam, a drunken farmer, riding home on his grey mare Maggie. As he passed Kirk Alloway, the church seemed to be ablaze. Dancing round the flames was a group of warlocks and witches, with the Devil himself playing the bagpipes. The astonished Tam saw that among the hag-like witches was one young and beautiful. Her name was Nannie, and she wore only a 'cutty sark', a short shirt of Paisley linen. The farmer was bewitched and, as her dancing became wilder, in his excitement:

> Tam tint his reason a'thegither
> An' roars out 'Weel done, Cutty Sark'.
> An' in an instant a' was dark.

Pursued by the witches, Tam fled for his life to the bridge over the Doon, for he knew that they could not cross running water. Nannie was faster than the others and, as the mare galloped over the bridge, she seized it by the tail, which came off in her hand.

That is why in the *Cutty Sark*'s figurehead Nannie's left arm is extended, with clutching fingers. In her racing days, after a fast passage, the apprentices would sometimes make a mare's tail from old rope, rubbed with grey paint, to put in her hand.

BELOW:
The original 'cutty sark' emblem, made of metal and worn on the mainmast to distinguish the ship when in port. Rescued from the dismasting in 1916, it subsequently turned up in a London saleroom in 1960, and was bought for the ship after a tip-off only 30 minutes before the start of the auction.

LEFT:
The legend of the 'cutty sark'. Tam O'Shanter rides for his life with Nannie in pursuit. Once across the bridge, Tam will be safe, for witches cannot cross running water.

The Tea Trade and Beyond

ABOVE:
A painting by Francis Smitheman of the Cutty Sark *leaving Shanghai on 17 June 1872. Thermopylae, with whom she was to race, is in the background.*

Between her maiden voyage in January 1870 and 1877, the year when she carried her last cargo of tea, *Cutty Sark* put up some sterling performances, without ever scaling great heights. Only once did she challenge *Thermopylae* when the ships loaded at the same time, and the race proved something of an anti-climax.

Both ships left Shanghai on 17 June 1872, bound for London via the Cape. Great excitement overtook both crews as they caught glimpses of the other ship while sailing down towards Borneo. In mid-July they were level, but then the *Cutty Sark* caught the south-east trades and really started to fly. The logbooks reveal that she was 400 miles ahead when disaster struck. On 14 August, a heavy sea tore the rudder from its bolts. The owner's brother, Robert Willis, begged Captain Moodie to head for South Africa. Moodie would have none of it and the two men nearly came to blows. The Captain prevailed. In five days hove to in heavy, rolling seas, the crew constructed and positioned a temporary rudder – with help from two stowaways

The Tea Trade and Beyond

LEFT:
Crew members on deck during a voyage. The man is repairing a sail. The jobs of a ship's boy would have included holystoning decks, polishing brass, being 'peggy' for the AB's and learning about life at sea.

BELOW:
Captain George Moodie, first Master of the Cutty Sark, *in formal and less formal poses. A determined and cautious Scot, it was his intolerance of anything less than perfection in the building materials which broke the firm of builders but probably accounts for the fact that the* Cutty Sark *survives today.*

– one a carpenter and one a blacksmith! By this time *Thermopylae* was some 500 miles ahead.

The *Cutty Sark* steered surprisingly well with its temporary rudder, but speed had to be kept down when the winds were on the beam. In the South Atlantic the jury rudder was once more hauled on deck for emergency repairs. But there were no problems this time and after two days the ship was under way again. On 18 October she passed Gravesend, 122 days out from Shanghai. *Thermopylae* had won the race in 115 days, but the glory went to Captain Moodie.

This did not stop the Captain from resigning, because of the row he had had with Robert Willis. Despite the pleas of his owner, Moodie began a new career in steam ships.

The days of the tea clipper were numbered. Passing through the newly-opened Suez Canal, steamships took half the time, were more reliable and undercut the sailing ships on price.

So it was that after eight years in the tea trade, *Cutty Sark* was forced to seek cargoes where she could get them, to Sydney with general cargo, on to Shanghai with coal then back to Sydney for more coal. In 1880, *Cutty Sark* carried wool from Melbourne to New York, a pointer to where her future lay.

In that same year, her lower masts and yards were reduced and her skysail and stunsails done away with, improving her performance in heavy seas.

Until 1882, she shuttled from east to west with all sorts of cargoes: jute for New York, oil for Java, buffalo horns for London, scrap iron for Shanghai. Yet, even deeply laden, she could still sail over 1,000 miles (1,600 km) in three days. Despite this, owner John Willis thought that the *Cutty Sark*'s racing days were over. How wrong he was to be proved.

A Hell Voyage

The *Cutty Sark* was known as a lucky ship, but she was not always a happy one. One voyage in 1880 was particularly ill-starred.

The seeds of tragedy were there from the outset. On 4 June, under the well-respected Captain Wallace, *Cutty Sark* sailed from London to South Wales to pick up a cargo of coal for the American Navy off Japan. The mate, Smith, was a hard-bitten Scot with a nasty reputation. Soon sensing this, most of the crew disappeared at Penarth so the 28 men that sailed for the East were a scratch bunch of ill-assorted nationalities.

At once one of them, a real sea-croaker (moaner) began to prophesy all sorts of disasters ahead. Soon his prophesies began to come true. First there were storms. Then the mate crossed swords with a crew member, John Francis, after the sailor's hand was severely mangled in a block. The crew were up in arms and the Captain ordered Francis to apologize or fight. The two men fought, and an uneasy truce followed.

After a good run south, hurricane-force winds hit the ship for three days near the Cape. This was traumatic for the crew, but brought the best out of the *Cutty Sark* which ran 1,050 miles (1,700 km) in three days. In the Sunda Straits, the previous trouble flared again. After a dispute, Francis took an iron bar to the mate, Smith, who turned the tables and killed the sailor with a blow to the neck. The mate was ordered to his cabin for the rest of the passage and the ship suddenly became very silent.

At Anjer in Java, Wallace smuggled Smith aboard an American vessel, the *Colorada*, which welcomed the idea of a 'man-handler' aboard. But the crew, who wanted Smith brought to book for murder, noticed he was gone and refused to work until the mate was handed over. Nevertheless, Wallace decided to sail, using his apprentices and petty officers to man the sheets. No sooner was the ship out of port than they struck several days of flat calm, much to the delight of the sea-croaker.

The steamy heat, the stagnant calm and the crew's smouldering resentment proved a volatile cocktail. Captain Wallace came to realize that in allowing the mate to escape he had put his career in jeopardy. Something had to happen, and on the fourth day of calm it did. Wallace gave some advice to the helmsman, walked aft, stepped on to the stern rail and jumped overboard. No trace of him was ever found – but the sharks that circled round told the tale of the fate that had met a fine seaman.

David Cobb's painting recalls the incident in 1888 when Cutty Sark, *outward bound for Sydney, overhauled the mail steamer* Britannia.

Principal dimensions:

Length overall (extreme):
 280 feet (86m)
Beam:
 36 feet (11m)
Depth (moulded):
 22.5 feet (7m)
Gross tonnage:
 936 tons
Net tonnage:
 921 tons
Displacement at 20 feet draught:
 2,100 tons
Sail area:
 32,000 square feet (2,972m²)

Signal Flags:

Cutty Sark's letters in the
International Code: JKWS

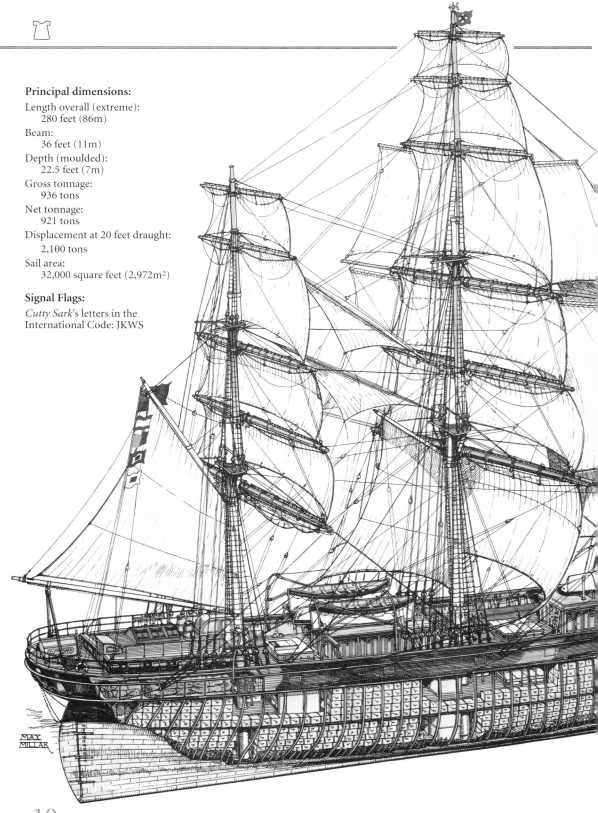

The Cutty Sark

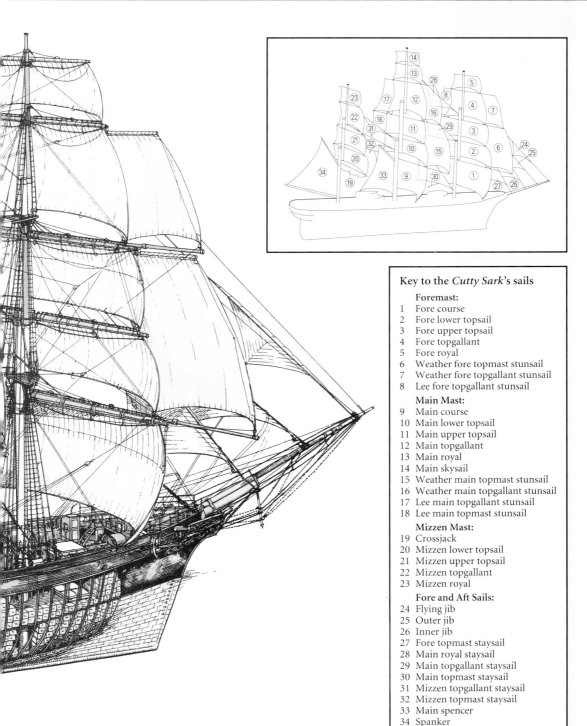

Key to the *Cutty Sark*'s sails

Foremast:
1. Fore course
2. Fore lower topsail
3. Fore upper topsail
4. Fore topgallant
5. Fore royal
6. Weather fore topmast stunsail
7. Weather fore topgallant stunsail
8. Lee fore topgallant stunsail

Main Mast:
9. Main course
10. Main lower topsail
11. Main upper topsail
12. Main topgallant
13. Main royal
14. Main skysail
15. Weather main topmast stunsail
16. Weather main topgallant stunsail
17. Lee main topgallant stunsail
18. Lee main topmast stunsail

Mizzen Mast:
19. Crossjack
20. Mizzen lower topsail
21. Mizzen upper topsail
22. Mizzen topgallant
23. Mizzen royal

Fore and Aft Sails:
24. Flying jib
25. Outer jib
26. Inner jib
27. Fore topmast staysail
28. Main royal staysail
29. Main topgallant staysail
30. Main topmast staysail
31. Mizzen topgallant staysail
32. Mizzen topmast staysail
33. Main spencer
34. Spanker

At Work in the Wool Trade

ABOVE:
Cutty Sark *loading wool bales at Circular Quay, Sydney.*

BELOW:
An Australian advertisement.

In 1883 *Cutty Sark*'s heyday was still ahead of her, although her owner 'White Hat' Willis didn't realize it. It was in that year that the ship entered the regular Australian wool trade, starting a long series of annual voyages, sailing out by the Cape and back by the Horn.

The clipper had been built for the 'Flying Fish' run to China, but it was in the rough weather of the Roaring Forties, the hardest ocean road of all, south of latitude 40°S, that she really proved her worth. Many sailors testified that they had never seen the *Cutty Sark* outrun by any ship, steam or sail.

Her first passage home with wool gave a foretaste of future achievements; from Newcastle, New South Wales, she made the Channel in 82 days.

Cutty Sark's finest hour was under Richard Woodget, Master 1885–95. Never have a ship and her master been more happily or successfully matched. Woodget was more than a fine seaman. He sensed the *Cutty Sark*'s amazing qualities of resilience and drove the ship to her limits in the vilest of weathers. Never afraid to go aloft himself, he dared to spread more canvas than lesser mortals would have done and so got the last quarter-knot of speed from the little

ship. In the same way, he drove his crew to the limits of their stamina but, like the vessel, they responded with an enthusiasm borne out of deep respect.

The combination proved unbeatable, and for the ten years from 1885 the *Cutty Sark* made the fastest passage home with wool, the first five of these beating *Thermopylae*, until her old rival was sold into the trans-Pacific rice trade.

> 'Captains do not like to admit that the Cutty Sark can sail, and yet not one of them can show that she has ever been beaten by any sailing vessel that has left London or Sydney about the same time.'
>
> RICHARD WOODGET, MASTER OF THE *CUTTY SARK* 1885–95.

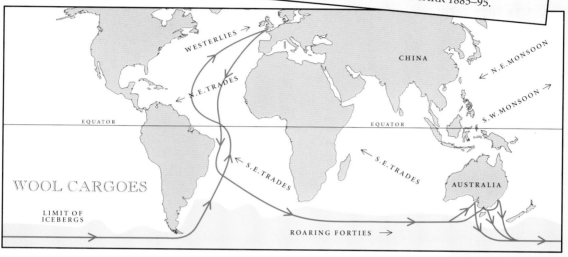

ABOVE:
The route to and from Australia taken by ships in the wool trade.

LEFT:
The crew of the Cutty Sark *with guests taken during the time when Captain Woodget was Master (1885–95). Woodget is seen in his tam o' shanter third from the right on the top row. Besides being an outstanding seaman, he was a keen amateur photographer and many of his pictures survive in the* Cutty Sark's *archives.*

Decline and Restoration

RIGHT:
A model of the Cutty Sark *in her days as the Portuguese* Ferreira. *In 1916 she was dismasted in a gale off the Cape and subsequently re-rigged with reduced sail as a barquentine.*

BELOW:
A dilapidated Cutty Sark *as the* Maria do Amparo *in Surrey Docks in 1922.*

Despite her remarkable achievements, by 1895 the *Cutty Sark* was no longer making money. Her cargo of wool that year was her biggest but her last under the British flag.

She was sold to the Portuguese, officially as the *Ferreira*, but known always to her crews as the *Pequina Camisola* ('little shirt'). So began 27 years of faithful but unglamorous service between Portugal and its colonies of South America and East Africa.

Her regular yearly round became: Oporto – Rio – New Orleans – Lisbon. She lost her rudder twice in storms, in 1909 and in 1915. In 1916 she was dismasted in the Indian Ocean and towed into Cape Town for repairs. Because new

spars were hard to come by during the First World War, she was re-rigged as a barquentine.

In 1920 she was sold again to become the *Maria do Amparo* and in 1922 was refitted in London. On her way home to Portugal, she was driven into Falmouth by a Channel gale, where she was seen by an old admirer. A new chapter of her life was about to begin.

A Cornish mariner, Captain Wilfred Dowman, had fallen in love with the *Cutty Sark* in 1894 when, as an apprentice seaman, he had seen her sail past at full speed. Seeing the ship in Falmouth, 28 years later, he decided to buy her back from Portugal. So, for the sum of £3,750, the *Cutty Sark* regained the Red Ensign and her old name. After all her wanderings, she had come home.

By 1924 the ship was re-rigged and restored as a clipper. On the death of Captain Dowman, his widow presented the *Cutty Sark* to the Incorporated Thames Nautical Training College at Greenhithe, where in 1938 she joined *HMS Worcester* as a training ship. After the Second World War, the college acquired a larger steel-built ship and no longer required the *Cutty Sark*.

After lengthy discussions about her future, the *Cutty Sark* was moved to a mooring off Greenwich so she could be exhibited during the Festival of Britain in 1951, and the London County Council provided £4,000 to enable her to be docked for survey and generally made presentable. Later that year HRH The Duke of Edinburgh brought together a

Decline and Restoration

committee to raise funds to build a new dock at Greenwich, and restore the ship for public exhibition.

She was now gifted to the newly-formed Cutty Sark Society. Work on restoration and dock construction began early in 1954. In December the *Cutty Sark* was moved to Greenwich and berthed where she now lies. By June 1957 restoration work on the hull and rigging was complete and the *Cutty Sark* was formally opened to the public by HM The Queen.

In preparing the *Cutty Sark* for exhibition, the object was to restore her as completely as possible to her original condition as a tea clipper of the early 1870s. We see her today as she might have been between voyages in her home port of London, with all her standing rigging set up taut, and her running rigging still rove, but her sails unbent and stowed below in the sail locker.

Research to achieve this result was a very difficult proposition. The only plans to survive into the 20th century had been destroyed by fire in Glasgow soon after the Second World War.

LEFT:
Cadets crowd Cutty Sark's *stern during her time as a training ship with* HMS Worcester *at Greenhithe (on the Thames near Dartford). Her pristine condition suggests that this may have been in 1938 when she was first moored there after her final sea voyage.*

A set of general plans came from the son of the chief draughtsman at Scott & Linton's yard, Mr John Rennie. The granddaughter of Hercules Linton loaned the original sail and rigging plan and sketches by her grandfather of the stern carving. The 90-year-old daughter of Henry Henderson, the *Cutty Sark*'s first ship's carpenter, lent her father's notebook, with dimensions of every mast and spar, and details of all the boats.

BELOW LEFT:
In 1954, the Cutty Sark *was manoeuvred into the dock where she is today.*

BELOW RIGHT:
The condition of the decks before her overhaul at Blackwall in 1954.

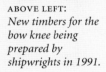

The Cutty Sark Today

ABOVE LEFT:
New timbers for the bow knee being prepared by shipwrights in 1991.

ABOVE RIGHT:
Shipwrights inspect work on the bow of the Cutty Sark, *part of the ongoing restoration programme that has been implemented by The Maritime Trust, now responsible for the ship.*

RIGHT:
The crew of the Cutty Sark *in their quarters in the forward deckhouse, where they lived in cramped conditions for the duration of a voyage. The crew, divided into two watches, worked at least 12 hours a day in good weather. In bad weather they would be on deck many more hours under cold, wet and dangerous conditions.*

Since the *Cutty Sark*'s permanent docking at Greenwich, visitors arrive by an entrance that was not there in the ship's heyday. The 'tween deck where we now board the ship was an area for the carriage of cargo, not a space for crew accommodation as one might suppose. The portholes which add to this impression date only from 1938, the start of the *Cutty Sark*'s days as a cadet training ship.

Most of the crew were housed at the fore end in the fo'c'sle, now used as a ship's store. The accommodation was divided from the hold by a watertight bulkhead and could only be entered by a small square hatch on the main deck. When fitted with bunks for 22 men it must have been a dark and dismal place, lit by only four small portholes on each side. Between 1870 and 1872, the crew's lot was improved by the addition of a second deckhouse providing better accommodation.

The after part of the 'tween deck, then crammed with cargo, now contains a display telling the history of the ship and the trades in which she sailed. Here on show is the ship's original figurehead and also important models of the ship. The model of the *Cutty Sark*, rigged as she was under the British flag, was constructed by Sir Maurice Denny and presented to the ship by his widow. It is probably the finest small model of a clipper ship ever made. It is interesting to contrast this with another model on display (see page 14) showing her rigged as the barquentine *Ferreira* during her later days under the Portuguese flag.

It was in the 'tween deck, when it was empty of cargo, that Captain Woodget patiently taught himself to ride a bicycle in the 1890s. His mount was a real old 'bone-shaker' with wooden frame and wooden wheels.

The Cutty Sark Today

LEFT:
The saloon, panelled in teak and bird's-eye maple, was used by the ship's officers as a chart room for navigation as well as a dining room.

BELOW:
A crew member sits on the main hatch and splices a rope. With 11 miles (18km) of rigging used to control the sails, this was an essential task to ensure the Cutty Sark's smooth operation.

The Cutty Sark Today

RIGHT:
A £2,000,000, ten-year restoration began in 1992.

BELOW:
Cutty Sark's figurehead during restoration.

BOTTOM:
Part of the fine display of ships' figureheads.

In the *Cutty Sark*'s lower hold, a cargo space in her working days, there was neither deck nor stairway. Now visitors can view here the finest collection of merchant ship figureheads in the country and see a video display showing archive film of life aboard the great steel sailing ships, taken in the 1920s with hand-held clockwork cameras.

On the maindeck (topdeck), looking aloft, one's eye is drawn to the towering mainmast, standing 152 feet (47m) above and the complexity of rigging, 11 miles (18km) of it, needed to control 34 sails. With all sail set, the *Cutty Sark* effectively developed 2,250 horse power, driving the ship forward at 17 knots (19.5 mph or 31.5 kph).

The crew accommodation here ranges from the comparatively palatial officers' quarters in the poop, to the more basic ratings' quarters in the forward deckhouse. It is amazing to think that a ship of this size could be sailed by only 28 people, eight of whom were boy apprentices.

The Cutty Sark Today

Standing by the wheel and gazing forward and upward, it is not difficult to imagine the ship at sea, as seen by the helmsman, roaring through the water at full speed, with the bow wave creaming away on either side and the great seas rolling up astern. One can almost feel the lift and heave of the reeling deck. This was life at sea in the great days of sail, this the strength and beauty of the sailing ship. Aboard the *Cutty Sark*, one can feel a bond of sympathy with John Masefield's 'Dauber' when he said:

*Ships and the sea;
there's nothing finer made.*

LEFT:
A view of the stern of the Cutty Sark *in her permanent berth at Greenwich. In the centre of the elaborate gilded carving known as 'gingerbread' is the Star of India badge. Below this is the motto 'Where there's a will is a way', a play on the name of the first owner John Willis.*

OVERLEAF:
Looking forward from the ship's wheel on the poop deck, from where the helmsman steered the ship. In the background are two crew members attending to the rigging around the mizzen mast.

The Cutty Sark Today